Muscles

by Justin McCory Martin

ISBN 978-0-545-24796-2

Photographs © 2010: cover: Getty Images/Bill Reitzel/Plush Studios; back cover top: iStockphoto/Dmitriy Shironosov; back cover bottom: iStockphoto/Nathan Jones; page 1: iStockphoto/Nathan Jones; page 2: ShutterStock, Inc./Bronwyn Photo; page 3: Getty Images/Dorling Kindersley; page 4: iStockphoto/Jacek Chabraszewski; page 5 left: iStockphoto/Jani Bryson; page 5 right: Photo Researchers, NY/BSIP; page 6: iStockphoto/Daniel Laflor; page 7: iStockphoto/Mark Fairey; page 8: iStockphoto/Sergey Mostovoy; page 9 left: Getty Images/Andy Crawford; page 9 right: Getty Images/Dorling Kindersley; page 10 left: iStockphoto/Leah-Anne Thompson; page 10 right: iStockphoto/Sebastian Kaulitzki; page 11 left: Peter Arnold Inc./MedicalRF/The Medical File; page 11 right: iStockphoto/Lesley Lister; page 12 left: Getty Images/Dorling Kindersley; page 12 right: Getty Images/Dorling Kindersley; page 13 top left: iStockphoto/Gerville Hall; page 13 top right: iStockphoto/Tomasz Marawski; page 13 bottom: iStockphoto/Chris Martinov; page 14: iStockphoto/Jacek Chabraszewski; page 15: iStockphoto/Jordan Shaw; page 16: Getty Images/Dorling Kindersley.

Photo research by Jenna Addesso; Design by Holly Grundon

12 11 10 9 8 7 6 5 4 3 2 1   10 11 12 13 14 15/0
Printed in the U.S.A.   40
First printing, September 2010

SCHOLASTIC INC.

NEW YORK • TORONTO • LONDON • AUCKLAND
SYDNEY • MEXICO CITY • NEW DELHI • HONG KONG

Make a muscle!

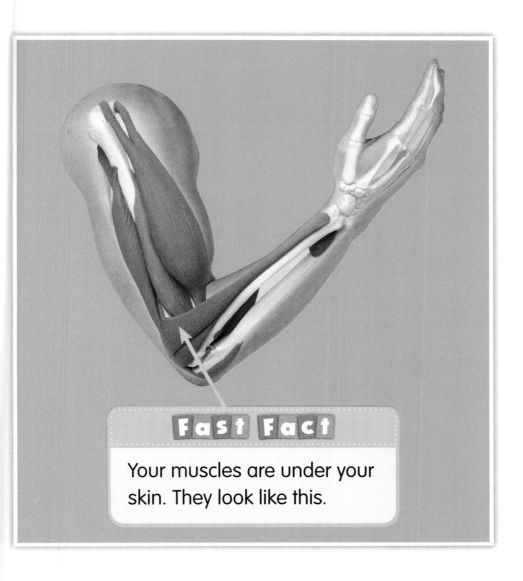

When you flex your arm,
muscles are **hard** at **work**!

Fast Fact

The **more** you move, the **more** muscles you use.

When you run,
muscles are **hard** at **work**.

Muscles are made of stretchy tissue. They look like this.

When you dance, muscles are **hard** at **work**.

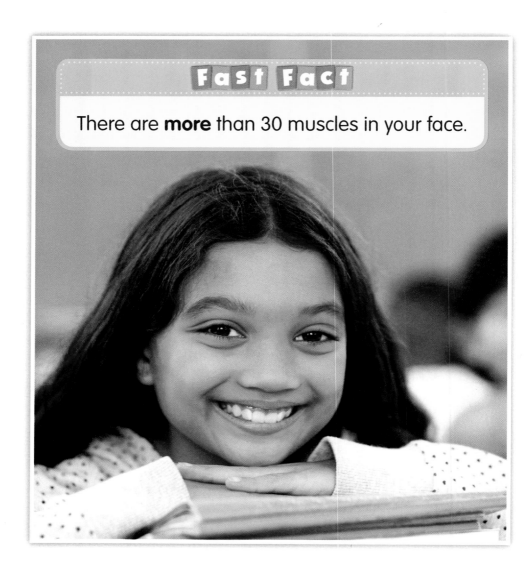

There are **more** than 30 muscles in your face.

When you smile,
muscles are **hard** at **work**.

You use **more** muscles to frown than to smile.

When you frown,
muscles are **hard** at **work**.

The tiny muscles in your eyes move **more** than 100,000 times each day!

When you look around, muscles are **hard** at **work**.

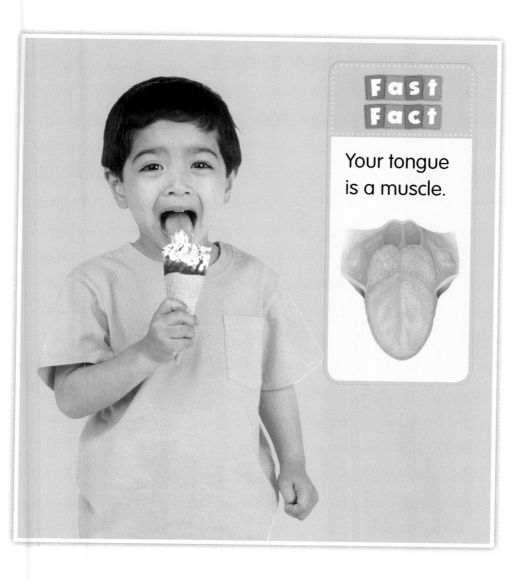

Fast Fact

Your tongue is a muscle.

When you lick an ice-cream cone, muscles are **hard** at **work**.

Fast Fact

Some muscles **work** all by themselves. You breathe without even thinking about it.

When you breathe, muscles are **hard** at **work**.

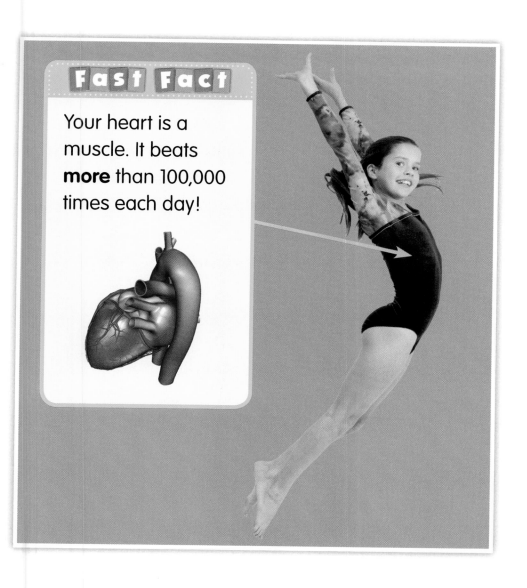

Your heart is a muscle. It beats **more** than 100,000 times each day!

When your heart beats, muscles are **hard** at **work**.

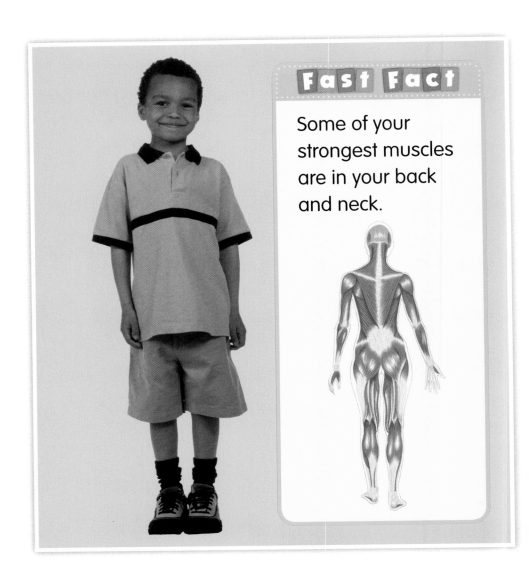

Some of your strongest muscles are in your back and neck.

Even **when** you stand still, muscles are **hard** at **work**.

karate

biking

skating

Some ways to make your muscles strong:
- karate
- biking
- skating
- running
- swimming

Thanks, muscles! Keep up the great **work**!

Sight Word Review

Point to each sight word. Then read it aloud.

more

work

hard

when

work

hard

when

more

Sight Word Fill-ins

Use one sight word from the box to finish each sentence.

hard	more
when	work

1 You have _____ than 600 muscles in your body.

2 Your muscles are always at _____ .

3 Doing 50 push-ups is _____ !

4 Even _____ you stand still, your muscles are working.

All About Muscles

Ask a grown-up to read this with you.

You need muscles to move. Without muscles, you could not walk or talk, sit or smile, run or jump.

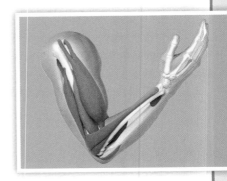

You have more than 600 muscles. Nearly half the weight of your body comes from muscles.

The strong muscles in your back help you stretch for the sky or touch your toes. The little muscles in your face let you smile, frown, or wriggle your nose.

You control the movement of some muscles. You can choose to move your arms to clap your hands together. Muscles that you control are called voluntary muscles. Involuntary muscles are muscles that work all by themselves. Your heart is an involuntary muscle. It beats without you even thinking about it. When you eat, your stomach muscles work on their own to digest food.

Muscles are stretchy, kind of like rubber bands. They are made of elastic tissue. You can make your muscles stronger by exercising. Do push-ups and sit-ups and watch your muscles grow stronger and stronger!